梁燕 編著

# 惹味咖喱

前言

四大文明古國之一的印度是咖喱的發源地，在食的文化領域風靡全世界，而各國如馬來西亞、印尼、泰國、緬甸等的咖喱有着本身不同的特色，有紅咖喱、黃咖喱、綠咖喱多種菜式。

初嘗咖喱菜餚的食客，會覺得味道香濃，能刺激食慾，而香味與辣味之間的比重，隨着每個人的口味而不同；鹹度、酸度、辣度、甜度的變化，隨着不同菜式的製作而變化無窮。

再者，配以各種香料，令咖喱更廣受歡迎，如白豆蔻、小茴香、丁香、香葉、酸子等。而食材選擇方面更加變化多端，肉類有羊、雞、牛、豬，海產可以用魚、蝦、蟹、貝殼類，蔬菜有西蘭花、椰菜花、蘑菇等，其中薯仔更是一個十分重要的角色。咖喱汁味道濃烈，配以麵包、薄餅、白飯等原味的食物，顯出其特色，受歡迎程度更能經得起考驗。

# 目錄

# 嘢囉角

## Deep-fried Curry Beef Dumplings

### 材料 | Ingredients

| | |
|---|---|
| 牛肉 150 克 | 150g beef |
| 薯仔 2 個 | 2 potatoes |
| 洋葱 1 個 | 1 onion |
| 青豆 1/2 杯 | 1/2 cup green peas |
| 春卷皮 1 包 | 1 pack spring roll wrappers |
| 牛油 50 克 | 50g butter |

### 醃料 | Marinade

| | |
|---|---|
| 咖喱醬 1 1/2 茶匙 | 1 1/2 tsps curry paste |
| 麻油 1 茶匙 | 1 tsp sesame oil |
| 生粉 1 茶匙 | 1 tsp caltrop starch |
| 咖喱粉 3/4 茶匙 | 3/4 tsp curry powder |
| 胡椒粉 1/2 茶匙 | 1/2 tsp ground white pepper |
| 酒 1/2 茶匙 | 1/2 tsp wine |
| 鹽 1/4 茶匙 | 1/4 tsp salt |
| 糖 1/4 茶匙 | 1/4 tsp sugar |
| 油 1 茶匙（後下） | 1 tsp oil (added lastly) |

## 粉漿 | Flour paste

麵粉 1/2 湯匙
水 1/2 湯匙
1/2 tbsp flour
1/2 tbsp water

餅皮裁成 14cm×7cm 平行四邊形
Cut the wrappers into parallelograms as below

## 做法 | Method

1. 牛肉洗淨，瀝乾水分，剁碎，加醃料拌勻，最後加油拌勻。

2. 薯仔洗淨，去皮，下水中煮至腍，壓成茸，加牛油拌勻。

3. 洋葱去衣，洗淨，切幼粒，下油鑊爆香備用。

4. 青豆洗淨，下鹽水中灼水，瀝乾水分。

5. 燒熱油鑊，下牛肉、洋葱、薯茸、青豆拌炒勻成餡料。

6. 春卷裁成 14 厘米 ×7 厘米平行四邊形，包入餡料，摺疊成三角形，用粉漿封口，下油鑊中炸至金黃即成。

1. Rinse and drain beef. Chop into mash and mix with the marinade. Add oil at last and mix well.

2. Rinse potatoes. Peel and cook in water until tender. Mash and mix with butter.

3. Skin onion. Rinse and cut into fine dice. Stir-fry with oil until fragrant. Set aside.

4. Rinse green peas and scald them in salty water. Drain.

5. Heat oil in a wok. Put in beef, onion, potato mash and green peas. Stir-fry well to make the filling.

6. Cut spring roll wrappers into parallelograms of base length 14 cm and perpendicular height 7 cm as shown in the picture. Wrap in the filling and fold into triangular shapes. Bind the ends with the flour paste. Deep-fry in oil until golden brown. Serve.

---

小貼士 | Tips

嚤囉角可以預早一晚包好，放雪櫃中保存，油炸時才取出。

# 咖喱雞柳串

## Curry Chicken Skewers

### 材料 | Ingredients

雞肉（去骨）500 克
紅椒 1 個
青椒 1 個

500g boned chicken
1 red bell pepper
1 green bell pepper

2~4 人
2~4 persons

15~20 分鐘
15~20 minutes

## 醃料 | Marinade

| | |
|---|---|
| 蒜茸 2 粒量 | 2 cloves garlic (grated) |
| 咖喱醬 2 湯匙 | 2 tbsps curry paste |
| 生粉 3 茶匙 | 3 tsps caltrop starch |
| 雞粉 2 茶匙 | 2 tsps chicken powder |
| 香茅粉 1 茶匙 | 1 tsp lemongrass powder |
| 胡椒粉 1/2 茶匙 | 1/2 tsp ground white pepper |
| 糖 1/2 茶匙 | 1/2 tsp sugar |

## 調味料 | Seasonings

咖喱醬 2 湯匙
2 tbsps curry paste

**小貼士 | Tips**

此菜式可加入其他蔬菜，例如菠蘿、青瓜等。

## 做法 | Method

1. 雞肉洗淨，瀝乾水分，切片，用刀背打鬆肉質，加醃料醃 20 分鐘。
2. 紅椒、青椒洗淨，去籽，切角。
3. 將雞片、紅椒和青椒相間地穿在竹籤上。
4. 燒熱油鑊，將雞串炸至金黃色，塗上或蘸上咖喱醬便可供食。

1. Rinse chicken. Wipe dry and cut into slices. Tap with the back of a knife to tenderize it. Marinate for 20 minutes.
2. Rinse red and green bell peppers. Seed and cut into wedges.
3. Skewer chicken slices, red and green bell pepper in order with bamboo skewers.
4. Heat oil in a wok. Deep-fry the skewers until golden brown. Rub over curry paste or serve with curry paste as a dip.

## 咖喱長通粉沙律

### Curry Penne Salad

◯◯◯ 材料 | Ingredients

長通粉 250 克
粟米粒 250 克
菠蘿 4 片
腸仔 4 條
火腿 2 片
子薑適量
鹽 1 茶匙

250g penne
250g corn kernels
4 slices pineapple
4 sausages
2 slices ham
young ginger
1 tsp salt

## ⟨⟨⟩⟩ 調味料 | Seasoning

| | |
|---|---|
| 白汁 220 克 | 220g white sauce |
| 茄汁 3 湯匙 | 3 tbsps ketchup |
| 咖喱粉 2 湯匙 | 2 tbsps curry powder |
| 鹽 1 茶匙 | 1 tsp salt |
| 胡椒粉 1/2 茶匙 | 1/2 tsp ground white pepper |
| 拔蘭地酒數滴 | several drops brandy |

## ⟨⟨⟩⟩ 做法 | Method

1. 燒滾水，下鹽，加入長通粉煮至腍，用凍開水過冷河。
2. 腸仔、火腿洗淨，與菠蘿、子薑同切幼粒。
3. 將調味料拌勻，加入長通粉和其他材料再拌勻，放入雪櫃冷凍便可供食。

1. Bring water to the boil and add in salt. Put in penne and cook until tender. Rinse in cold drinking water and drain.
2. Rinse sausages and ham. Finely dice together with pineapple and young ginger.
3. Mix seasoning. Add into the penne and other ingredients. Mix well. Refrigerate and serve.

---

小貼士 | Tips

長通粉選用較短的一種，煮腍後切成粒狀，如菠蘿粒、腸仔粒般大小，看起來比較美觀。

# 咖喱雞雜果沙律

## Curry Chicken Salad with Assorted Fruits

### 材料 | Ingredients

雞髀肉 1 塊
西芹 100 克
罐頭雜果 1 杯
菠蘿 2 片

1 piece boned chicken leg
100g celery
1 cup canned assorted fruits
2 slices pineapple

4~6 人
4~6 persons

25~30 分鐘
25~30 minutes

### ⟨⟨⟩ 醃料 | Marinade

鹽 1 茶匙
薑汁 1 茶匙
酒 1 茶匙

1 tsp salt
1 tsp ginger juice
1 tsp wine

### ⟨⟨⟩ 沙律醬料 | Salad dressing

沙律醬 6 湯匙
咖喱粉 1 湯匙
鹽 1/4 茶匙

6 tbsps salad dressing
1 tbsp curry powder
1/4 tsp salt

### ⟨⟨⟩ 做法 | Method

1. 雞髀肉洗淨,瀝乾水分,加醃料醃 30 分鐘,隔水蒸約 12 分鐘至熟,待涼透後切粒。
2. 西芹洗淨,切小粒,汆水後瀝乾水分備用。
3. 菠蘿切粒。
4. 雞肉、西芹、雜果同放雪櫃中冷凍。
5. 拌勻沙律醬料,加入雞肉、西芹、雜果、菠蘿同拌勻,即可食用。

1. Rinse chicken leg flesh and drain. Marinate for 30 minutes. Steam for about 12 minutes until done. Set aside to let cool and dice.
2. Rinse celery and cut into small dice. Scald and drain.
3. Dice pineapple.
4. Refrigerate chicken, celery and assorted fruits together.
5. Mix the salad dressing. Add chicken, celery, assorted fruits and pineapple. Mix well and serve.

---

小貼士 | Tips

咖喱多加了也不會太辣,味道反而會更香。

## 泰式咖喱免治豬肉脆蝦片

Thai Curry Pork with Prawn Chips

### ◯◯◯ 材料 | Ingredients

免治豬肉 200 克
紅辣椒碎 1 隻量
蒜茸 1 茶匙
香茅碎 1 茶匙
蝦片或餅乾酌量
芫荽酌量

200g minced pork
1 chopped red chilli
1 tsp grated garlic
1 tsp chopped lemongrass
prawn chips or biscuits
coriander

◯◯◯ 調味料 | Seasoning

| | |
|---|---|
| 椰汁 240 毫升 | 240ml coconut milk |
| 魚露 2 湯匙 | 2 tbsps fish sauce |
| 黃糖 2 湯匙 | 2 tbsps brown sugar |
| 幼粒花生醬 3 茶匙 | 3 tsps fine peanut butter |
| 咖喱醬 2 茶匙 | 2 tsps curry paste |
| 辣椒膏 1 1/2 茶匙 | 1 1/2 tsps chilli paste |

◯◯◯ 做法 | Method

1. 燒熱油鑊，下油爆香蒜茸、香茅、紅辣椒、免治豬肉，煮約 2 分鐘。
2. 加調味料煮至汁液濃稠。
3. 將免治豬肉連汁放在蝦片或餅乾上，以芫荽裝飾即可。

1. Heat oil in a wok. Stir-fry grated garlic, chopped lemongrass, red chilli and minced pork for about 2 minutes until fragrant.
2. Add seasoning and cook until the sauce thickens.
3. Put the minced pork and sauce on top of each prawn chip or biscuit. Garnish with coriander. Serve.

小貼士 | Tips

如喜歡香滑感覺宜用幼粒花生醬，要有口感則可改用粗粒花生醬。免治豬肉可以雞肉或牛肉代替。

# 咖喱錦鹵雲吞

## Deep-fried Chicken Wantons with Curry Vegetables

### 材料 | Ingredients

| | |
|---|---|
| 雲吞皮 300 克 | 300g wanton wrappers |
| 雞肉 150 克 | 150g boned chicken |
| 罐裝菠蘿 4 片 | 4 slices canned pineapple |
| 番茄 2 個 | 2 tomatoes |
| 青瓜 1 個 | 1 cucumber |
| 葱 3 棵 | 3 stalks spring onion |
| 青椒 1 個 | 1 green bell pepper |
| 紅燈籠椒 1 個 | 1 red bell pepper |
| 雞蛋 1 隻 | 1 egg |
| 蛋液適量 | whisked egg |

### 醃料 | Marinade

| | |
|---|---|
| 咖喱粉 2 湯匙 | 2 tbsps curry powder |
| 糖 1/2 茶匙 | 1/2 tsp sugar |
| 鹽、生抽各 1 茶匙 | 1 tsp salt and light soy sauce |
| 生粉 1 茶匙 | 1 tsp caltrop starch |

### 芡汁 | Thickening

| | |
|---|---|
| 生粉 2 茶匙 | 2 tsps corn starch |
| 水 4 湯匙 | 4 tbsps water |

6~8 人
6~8 persons

45~50 分鐘
45~50 minutes

## 汁料 | Sauce

| | |
|---|---|
| 糖 4 湯匙 | 4 tbsps sugar |
| 茄汁 2 湯匙 | 2 tbsps ketchup |
| 酸子汁 3 湯匙（酸子 3 粒 | 3 tbsps tamarind juice |
| 用熱水 1/2 杯煮至出味） | (cook 3 tamarinds with 1/2 cup |
| 咖喱醬 3 茶匙 | of hot water until flavourful) |
| 酒 2 茶匙 | 3 tsps curry paste |
| 麻油、鹽各 1 茶匙 | 2 tsps wine, 1 cup water |
| 生粉 1 茶匙 | 1 tsp sesame oil, 1 tsp salt |
| 水 1 杯 | 1 tsp caltrop starch |

## 做法 | Method

**小貼士 | Tips**

青瓜洗淨後，切去頭尾，用來磨頂部至白色泡沫出現，可去除青瓜的澀味。

1. 雞肉洗淨，切粒剁成茸，加醃料醃 30 分鐘，加雞蛋拌勻備用。
2. 青瓜切去頭尾磨蕊汁，洗淨後切滾刀塊；菠蘿切扇塊；番茄、青紅椒洗淨切塊。
3. 部分葱白切度備用，其餘切幼粒，加入雞肉茸中拌勻。
4. 雲吞皮包入雞肉茸，成雲吞，用蛋液封口；燒熱油鑊，下油炸至金黃色，瀝乾油分，上碟。
5. 下油爆炒青瓜、番茄，調入鹽少許，盛起。
6. 下油爆香咖喱、葱白、青、紅椒，倒入汁料煮滾，加入生粉水勾芡，下青瓜、番茄、菠蘿拌勻，即可盛起與雲吞同吃。

1. Rinse boned chicken and cut into fine dice. Chop into mash and marinate for 30 minutes. Mix with egg and set aside.
2. Cut ends from cucumber and rub out bitter juice at both ends. Rinse and cut into wedges by rolling the cucumber. Cut pineapple into fan-shaped pieces. Rinse tomatoes and red and green peppers. Cut them into pieces.
3. Section some of the white part of spring onion. Cut the remaining into fine dice and mix with the chopped chicken.
4. Wrap the chicken mash with wanton wrappers and bind the ends with whisked egg. Heat oil in a wok and deep-fry the wantons until golden brown. Drain and put onto a plate.
5. Heat oil in a wok. Stir-fry cucumber and tomatoes quickly. Add a little salt and stir-fry well. Dish up.
6. Heat oil in a wok. Stir-fry curry, white part of spring onion, green and red peppers until fragrant. Pour in the sauce and bring to the boil. Add the thickening to thicken the sauce. Put in cucumber, tomatoes and pineapple. Mix well and serve with the wantons.

4~6 人
4~6 persons

30~35 分鐘
30~35minutes

# 咖喱雞肉薯仔卷

## Deep-fried Curry Chicken and Potatoes Rolls

### 材料 | Ingredients

薯仔 120 克
免治雞肉 120 克
洋葱 1 個
春卷皮 10 塊
咖喱粉 3 湯匙
豆蔻粉 1/2 茶匙

120g potatoes
120g minced chicken
1 onion
10 spring roll wrappers
3 tbsps curry powder
1/2 tsp ground nutmeg

## 醃料 | Marinade

生抽 3 茶匙　　　　　3 tsps light soy sauce
糖 1 茶匙　　　　　　1 tsp sugar
生粉 1 茶匙　　　　　1 tsp caltrop starch

## 做法 | Method

1. 薯仔洗淨，下沸水中焓熟後去皮壓成茸；洋葱洗淨，去衣切粒。

2. 免治雞肉加醃料醃 20 分鐘。

3. 燒熱油鑊，爆香洋葱，加免治雞肉炒至乾身，加入咖喱粉炒香，拌入薯茸，拌成餡料，稍待涼。

4. 用春卷皮包入餡料，捲成長條形，以少許水封口。

5. 燒熱油鑊，將雞肉卷放油中炸至金黃色，盛起，灑上豆蔻粉即成。

1. Rinse potatoes. Blanch in boiling water until done. Peel and mash. Rinse onion. Skin and cut into dice.

2. Marinate minced chicken for 20 minutes.

3. Heat oil in a wok. Stir-fry onion until fragrant. Add minced chicken and stir-fry until dry. Put in curry powder and stir-fry until fragrant. Mix in mashed potatoes. This makes the filling. Set aside to let cool.

4. Wrap the filling with spring roll wrappers. Roll into long strips and bind the ends with a little water.

5. Heat oil in a wok. Deep-fry the chicken rolls until golden brown and drain. Sprinkle over ground nutmeg and serve.

### 小貼士 | Tips

炸食物時，不可炸至深啡色，因撈起時，食物在油中會繼續炸，顏色會繼續加深。

# 越式酸辣咖喱魚湯

## Hot and Sour Curry Fish Soup in Vietnamese Style

### ⫝̸ 材料 | Ingredients

| | |
|---|---|
| 大眼魚 2 條 | 2 big-eye fish |
| 菠蘿 4 片 | 4 slices pineapple |
| 番茄 2 個 | 2 tomatoes |
| 洋葱茸 1 個量 | 1 onion (chopped) |
| 乾葱 3 粒 | 3 shallots |
| 咖喱醬 2 茶匙 | 2 tsps curry paste |
| 辣椒膏 1 茶匙 | 1 tsp chilli paste |
| 水約 5 杯 | approx. 5 cups water |

2~4 人
2~4 persons

20~25 分鐘
20~25 minutes

## 醃料 | Marinade

酒 2 茶匙
胡椒粉 1 茶匙
鹽 1/2 茶匙

2 tsps wine
1 tsp ground white pepper
1/2 tsp salt

## 調味料 | Seasoning

糖 1 茶匙
鹽 1/2 茶匙
胡椒粉少許
酸子 6 粒（以 1/2 杯水煮成酸汁）
水 1/2 杯

1 tsp sugar, 1/2 tsp salt
ground white pepper
6 tamarinds (cooked in 1/2 cup
of water into tamarind juice)
1/2 cup water

## 做法 | Method

1. 大眼魚劏洗淨，瀝乾水分，加入醃料醃 15 分鐘。
2. 番茄洗淨，切塊；菠蘿切塊；乾葱洗淨，去衣，略拍。
3. 燒熱油鑊，下魚煎至金黃，加水煮至魚湯變成奶白色，備用。
4. 起油鑊，加乾葱爆香，棄去，下咖喱、辣椒膏、洋葱茸、番茄炒勻，放入魚湯內，下調味料和菠蘿，煮滾後盛起即成。

1. Gut and rinse big-eye fish. Drain and marinate for 15 minutes.
2. Rinse tomatoes and cut into pieces. Cut pineapple into pieces. Rinse shallots. Skin and pat lightly.
3. Heat oil in a wok. Fry the fish until golden. Add in water and cook until the fish soup is milky white. Set aside.
4. Heat oil in a wok. Stir-fry the shallots until fragrant and remove them. Put in curry, chilli paste, chopped onion and tomatoes. Stir-fry well. Pour the mixture into the fish soup. Add seasoning and pineapple. Bring to the boil and serve.

小貼士 | Tips

魚煎香後，在熄火前即加水，魚湯的顏色才會變成奶白色。

# 番茄辣湯

## Hot Tomato Soup

### 材料 | Ingredients

番茄 500 克
洋葱 1 個
香葉 1 片
蒜肉 1 粒
雞湯 1 1/2 杯
鮮奶 1 杯
牛油 2 湯匙

500g tomatoes
1 onion
1 bay leave
1 clove skinned garlic
1 1/2 cups chicken stock
1 cup fresh milk
2 tbsps butter

⃝⃝ 調味料 | Seasoning

咖喱粉 1 茶匙       1 tsp curry powder
鹽 1/2 茶匙       1/2 tsp salt
胡椒粉 1/2 茶匙       1/2 tsp ground white pepper
辣汁 1/4 茶匙       1/4 tsp chilli sauce

⃝⃝ 做法 | Method

1. 番茄洗淨，切粗粒；洋葱去衣，洗淨，切幼粒。
2. 起油鑊，煮熔牛油，下洋葱炒軟，盛起。
3. 下油爆香蒜肉，加入洋葱、番茄、雞湯、調味料煮片刻，放入香葉煮約 10 分鐘。
4. 加入鮮奶，拌勻即可供食。

1. Rinse tomatoes and cut into large dice. Skin onion. Rinse and cut into fine dice.
2. Heat oil in a wok. Add butter and cook until molten. Add onion and stir-fry until tender. Dish up.
3. Heat oil in a wok. Stir-fry garlic until fragrant. Put in onion, tomatoes, chicken stock and seasoning. Cook for a while. Add bay leave and cook for about 10 minutes.
4. Add fresh milk. Mix well and serve.

小貼士 | Tips
鮮奶最後才下會令番茄辣湯更幼滑。

# 咖喱南瓜黃豆湯

Curry Pumpkin and Soybeans Soup

## 材料 | Ingredients

| | |
|---|---|
| 南瓜 300 克 | 300g pumpkin |
| 黃豆 200 克 | 200g soybeans |
| 洋葱 1 個 | 1 onion |
| 西芹 2 條 | 2 sprigs celery |
| 桂皮 1 段 | 1 section cinnamon |
| 丁香 2 粒 | 2 cloves |
| 八角 1 粒 | 1 star aniseed |
| 咖喱粉 2 湯匙 | 2 tbsps curry powder |
| 黃薑粉 1 湯匙 | 1 tbsp ground turmeric |
| 水 6 碗 | 6 bowls water |

6~8 人
6~8 persons

40~45 分鐘
40~45 minutes

## ⚭ 調味料 | Seasoning

魚露適量
fish sauce

## ⚭ 做法 | Method

1. 黃豆洗淨，瀝乾水分；下沸水中煮至腍，壓成茸。
2. 南瓜洗淨，去籽，切塊。
3. 洋葱洗淨，去衣切塊；西芹洗淨，切絲。
4. 水煮沸，加入洋葱、丁香、八角、桂皮，以中火煮 20 分鐘，濾出清湯備用。
5. 燒熱油鑊，爆香咖喱粉、黃薑粉，加入黃豆茸、南瓜，再放入清湯，慢火煮熔，將南瓜壓成茸，加入西芹和魚露即成。

1. Rinse soybeans and drain. Boil in boiling water until tender and mash.
2. Rinse pumpkin. Seed and cut into pieces.
3. Rinse onion. Skin and cut into pieces. Rinse celery and shred.
4. Bring water to the boil. Add onion, cloves, aniseed and cinnamon. Cook over medium heat for 20 minutes. Drain the soup and set aside.
5. Heat oil in a wok. Stir-fry curry powder and ground turmeric until fragrant. Add mashed soybean and pumpkin. Pour in the soup. Cook over low heat until dissolved. Mash the pumpkin. Add celery and fish sauce and serve.

2~4 人
2~4 persons

30~35 分鐘
30~35 minutes

青咖喱雞湯

Green Curry Chicken Soup

## 材料 | Ingredients

| | |
|---|---|
| 雞件 300 克 | 300g chicken (cut into pieces) |
| 茄子 1 條 | 1 eggplant |
| 青辣椒 1 隻 | 1 green chilli |
| 紅辣椒 1 隻 | 1 red chilli |
| 椰汁 400 毫升 | 400ml coconut milk |
| 綠咖喱醬 3 湯匙 | 3 tbsps green curry paste |
| 魚露 2 湯匙 | 2 tbsps fish sauce |
| 水 6 碗 | 6 bowls water |

## 做法 | Method

1. 雞件洗淨，瀝乾水分。
2. 茄子洗淨，切幼條；青、紅辣椒洗淨，切幼條。
3. 燒熱油鑊，加入雞件半煎炒，加入綠咖喱醬及茄子快炒。
4. 再加入椰汁和水，以中火煮 20 分鐘，加入青、紅辣椒，再下魚露調味，盛起即成。

1. Rinse chicken and drain.
2. Rinse eggplant and cut into thin strips. Rinse green and red chillies. Cut them into thin strips.
3. Heat oil in a wok. Add chicken and fry lightly. Put in green curry paste and eggplant. Stir-fry quickly.
4. Add in coconut milk and water again. Cook over medium heat for 20 minutes. Put in green and red chillies. Season with fish sauce. Serve.

---

小貼士 | Tips

茄子切開後要用水浸着才不會變黑。

## 日式咖喱雞

Japanese Curry Chicken

### ⃝⃝ 材料 | Ingredients

雞髀肉 2 塊
洋葱 2 個
蘋果 1/2 個
日式咖喱磚 100 克

2 pieces boned chicken leg
2 onions
1/2 apple
100g Japanese curry cube

2~4 人
2~4 persons

20~25 分鐘
20~25 minutes

## 醃料 | Marinade

生抽 1 茶匙
酒 1 茶匙
胡椒粉 1/2 茶匙
糖 1/2 茶匙

1 tsp light soy sauce
1 tsp wine
1/2 tsp ground white pepper
1/2 tsp sugar

## 調味料 | Seasoning

糖 1 1/2 茶匙
鹽 1/2 茶匙
水 1 杯

1 1/2 tsps sugar
1/2 tsp salt
1 cup water

## 做法 | Method

1. 雞髀肉洗淨，瀝乾水分，切粒，加醃料醃 30 分鐘。
2. 洋葱去衣，洗淨，切小方塊；蘋果去籽，切小方塊。
3. 燒熱油鑊，爆炒洋葱、雞肉、蘋果，炒至雞肉熟透，熄火，加入咖喱磚，慢慢拌勻至咖喱熔化。
4. 轉慢火煮片刻，加調味料拌勻即可。

1. Rinse chicken leg flesh. Drain and dice. Marinate for 30 minutes.
2. Skin onions. Rinse and cut into small squares. Seed apple and cut into small squares.
3. Heat oil in a wok. Stir-fry onions, chicken and apple together until the chicken is done. Remove from heat and add curry cube. Stir lightly until the curry molten.
4. Cook over low heat for a while. Put in seasoning and mix well. Serve.

---

小貼士 | Tips

蘋果去皮後會變成鏽色，所以未炒時，先要用水浸着。

# 咖喱腰果雞丁

## Curry Chicken Dice with Cashew Nuts

### 材料 | Ingredients

雞肉 200 克
洋葱 1 個
青椒 1 個
紅椒 1 個
冬菇 2 朵
炸腰果 50 克
咖喱粉 1 茶匙

200g chicken flesh
1 onion
1 green bell pepper
1 red bell pepper
2 dried black mushrooms
50g deep-fried cashew nuts
1 tsp curry powder

### 醃料 | Marinade

鹽 1 茶匙
酒 1 茶匙
胡椒粉 1/2 茶匙

1 tsp salt
1 tsp wine
1/2 tsp ground white pepper

### 調味料 | Seasoning

生抽 1 茶匙
糖 1/2 茶匙
水少許

1 tsp light soy sauce
1/2 tsp sugar
water

### 做法 | Method

1. 雞肉洗淨，瀝乾水分，切粒，加醃料醃 30 分鐘。
2. 青、紅椒洗淨，去籽、切角。洋葱去衣，洗淨，切塊。
3. 冬菇浸軟，去蒂，切粒。
4. 燒熱油鑊，爆香咖喱、洋葱、冬菇、雞肉，下調味料炒至雞肉熟透，加入青、紅椒、腰果拌勻即可。

1. Rinse chicken and drain. Dice and marinate for 30 minutes.
2. Rinse green and red bell peppers. Seed and cut into wedges. Skin onion. Rinse and cut into pieces.
3. Soak dried black mushrooms until soft. Remove the stalks and dice.
4. Heat oil in a wok. Stir-fry curry powder, onion, dried black mushrooms and chicken until fragrant. Add seasoning and stir-fry until the chicken is done. Put in green and red bell peppers and cashew nuts. Mix and serve.

### 小貼士 | Tips

腰果要最後下才可保持鬆脆。

# 焗葡國雞

## Baked Chicken in Portuguese Style

### ◯◯◯ 材料 | Ingredients

| | |
|---|---|
| 光雞 1 隻 | 1 gutted chicken |
| 薯仔 500 克 | 500g potatoes |
| 洋葱 2 個 | 2 onions |
| 蒜頭 3 粒 | 3 cloves garlic |
| 乾葱 2 粒 | 2 shallots |
| 花生 3 湯匙 | 3 tbsps peanuts |
| 淡奶 160 毫升 | 160ml evaporated milk |
| 椰汁 160 毫升 | 160ml coconut milk |
| 牛油 50 克 | 50g butter |
| 咖喱醬 1 湯匙 | 1 tbsp curry paste |
| 咖喱粉 2 茶匙 | 2 tsps curry powder |

6~8 人
6~8 persons

45~50 分鐘
45~50 minutes

### ⊂⊃ 醃料 | Marinade

糖 2 茶匙
生抽 2 茶匙
生粉 2 茶匙
鹽 1 1/2 茶匙

2 tsps sugar
2 tsps light soy sauce
2 tsps caltrop starch
1 1/2 tsps salt

### ⊂⊃ 調味料 | Seasoning

鹽 1 茶匙
水約 3 杯

1 tsp salt
approx. 3 cups water

### 小貼士 | Tips

洋葱一定要炒過才
會發出香味。

### ⊂⊃ 做法 | Method

1. 光雞洗淨，斬件，以醃料醃 20 分鐘。
2. 薯仔去皮，洗淨，切滾刀塊；燒熱油鑊，下油炸至金黃色，盛起。
3. 洋葱去衣，洗淨，切粗塊，下油鑊加鹽略炒。
4. 燒熱油鑊，爆香蒜頭、乾葱、花生和咖喱，加入雞件、薯仔拌勻，下調味料煮 20 分鐘。
5. 將洋葱回鑊，加入淡奶、椰汁，再煮約 10 分鐘至各材料變腍，放在焗盆內，鋪上牛油，放入已預熱的焗爐，以 180℃焗 10 分鐘至金黃色即成。

1. Rinse chicken. Chop up and marinate for 20 minutes.
2. Peel and rinse potatoes. Cut it into wedges by rolling the potatoes. Heat oil in a wok. Deep-fry the potatoes until golden brown and drain.
3. Skin onions and rinse. Cut them into thick pieces. Stir-fry them with oil and salt briefly.
4. Heat oil in a wok. Stir-fry garlic, shallots, peanuts and curry until fragrant. Add chicken and potatoes. Mix well. Put in the seasoning and cook for 20 minutes.
5. Transfer the onions back to the wok. Add evaporated milk and coconut milk. Cook for about 10 minutes until the ingredients are tender. Put them into a baking tray and spread butter on top. Bake in a preheated oven at 180 ℃ for 10 minutes until golden brown. Serve.

## 咖喱雞絲拌冷麵

### Cold Soba with Curry Chicken

⊕⊕ 材料 | Ingredients

冷麵 300 克
雞肉 240 克
青瓜絲 1 條量
甘筍絲 1/2 個量
火腿 2 片
鹽少許

300g cold soba
240g chicken flesh
1 cucumber (shredded)
1/2 carrot (shredded)
2 slices ham
salt

⟨⟨⟩⟩ 汁料 | Sauce

咖喱粉 2 湯匙　　　　　2 tbsps curry powder
魚露 1 湯匙　　　　　　1 tbsp fish sauce
花生醬適量　　　　　　peanut butter
沙律醬適量　　　　　　salad dressing

⟨⟨⟩⟩ 做法 | Method

1. 青瓜絲用鹽少許醃片刻，瀝乾水分。
2. 甘筍絲汆水備用。
3. 火腿以沸水川燙，切絲。
4. 雞肉用少許鹽醃片刻，隔水蒸熟，撕成絲狀。
5. 冷麵以凍開水沖洗，放入盆內，順序加入青瓜、甘筍、火腿、雞肉，淋上已拌勻的汁料，置雪櫃冷凍後可供食用。

1. Marinate cucumber shreds with a little salt for a while and drain.
2. Scald carrot shreds and set aside.
3. Scald ham with boiling water and cut into shreds.
4. Marinate chicken with a little salt for a while. Steam until done and tease into shreds.
5. Rinse cold soba with cold drinking water. Put into a tray and add cucumber, carrot, ham and chicken in order. Pour over the mixed sauce. Refrigerate until cold. Serve.

小貼士 | Tips
用手將雞肉撕成絲狀比用刀切絲更有口感。

香草咖喱雞

Curry Chicken with Thyme

### 材料 | Ingredients

雞件 400 克
薯仔 3 個
洋葱 1 個
椰汁 160 毫升
咖喱粉 2 湯匙

400g chicken (cut into pieces)
3 potatoes
1 onion
160ml coconut milk
2 tbsps curry powder

2~4 人
2~4 persons

25~30 分鐘
25~30 minutes

## ⟨⟨⟨ 調味料 | Seasoning

酒 1 湯匙
胡椒粉 1 茶匙
鹽 1 1/2 茶匙
百里香少許

1 tbsp wine
1 tsp ground white pepper
1 1/2 tsps salt
thyme

## ⟨⟨⟨ 做法 | Method

1. 雞件洗淨，瀝乾水分。
2. 燒熱油鑊，下雞件泡油，盛起備用。
3. 洋葱洗淨，去衣切塊；薯仔洗淨，去皮切件。
4. 下油爆香咖喱、洋葱、薯仔，將雞件回鑊，加入調味料和椰汁，以慢火煮 20 分鐘即可上碟。

1. Rinse chicken and drain.
2. Heat oil in a wok. Stir-fry chicken briefly and drain. Set aside.
3. Rinse onion. Skin and cut into pieces. Rinse potatoes. Peel and cut into pieces.
4. Heat oil in a wok. Stir-fry curry, onion and potatoes until fragrant. Add chicken, seasoning and coconut milk. Simmer over low heat for 20 minutes. Serve.

---

小貼士 | Tips

將雞件泡油可令肉汁鎖在雞肉內。

## 緬甸椰汁酸辣雞湯檬

### Pho in Burmaese Sour and Spicy Chicken Soup

#### 材料 | Ingredients

| | |
|---|---|
| 檬粉 400 克 | 400g pho (Vietnamese noodles) |
| 雞件 300 克 | 300g chicken (cut into pieces) |
| 椰汁 400 毫升 | 400ml coconut milk |
| 香茅段 3 枝 | 3 sections lemongrass |
| 乾葱 2 個 | 2 shallots |
| 洋葱 1 個 | 1 onion |
| 水 31/2 杯 | 31/2 cups water |

#### 醃料 | Marinade

酒 1 茶匙
鹽 1/2 茶匙
胡椒粉 1/2 茶匙

1 tsp wine
1/2 tsp salt
1/2 tsp ground white pepper

## ⊗⊗ 調味料 | Seasoning

魚露 1 湯匙
糖 2 茶匙
鹽 1 茶匙

1 tbsp fish sauce
2 tsps sugar
1 tsp salt

## ⊗⊗ 咖喱汁 | Curry sauce

咖喱醬 1 1/2 湯匙
辣椒膏 1 湯匙
蒜茸 3 茶匙
乾葱茸 3 茶匙
咖喱粉 2 茶匙
酸子 3 粒（用熱水 1/2 杯煮至出味）

1 1/2 tbsps curry paste
1 tbsp chilli paste
3 tsps grated garlic
3 tsps grated shallots
2 tsps curry powder
3 tamarinds (cooked with 1/2 cup
of hot water until flavourful)

## ⊗⊗ 做法 | Method

1. 雞件洗淨，加醃料醃 20 分鐘。
2. 乾葱、洋葱洗淨，去衣切片。
3. 燒熱油鑊，爆香咖喱汁材料，加入雞件、乾葱、洋葱、香茅、水和調味料煮滾，下椰汁和檬粉煮沸即成。

1. Rinse chicken. Marinate for 20 minutes.
2. Rinse shallots and onion. Skin and slice.
3. Heat oil in a wok. Stir-fry the ingredients of curry sauce until fragrant. Add chicken, shallots, onion, lemongrass, water and the seasoning. Bring to the boil. Put in coconut milk and pho. Bring to the boil and serve.

---

小貼士 | Tips
檬粉不要煮太久，否則太脹會影響口感。

# 咖喱雞醬
## Curry Chicken Paste

### 材料 | Ingredients

| | |
|---|---|
| 雞肉 300 克 | 300g chicken flesh |
| 青椒 1 個 | 1 green bell pepper |
| 紅椒 1 個 | 1 red bell pepper |
| 洋葱 1/2 個 | 1/2 onion |
| 葡萄乾 1/2 杯 | 1/2 cup raisins |
| 乾葱茸 3 粒量 | 3 shallots (grated) |
| 蒜茸 2 粒量 | 2 cloves garlic (grated) |
| 咖喱醬 2 湯匙 | 2 tbsps curry paste |
| 咖喱粉 2 茶匙 | 2 tsps curry powder |

2~4 人
2~4 persons

15~20 分鐘
15~20 minutes

40

## 醃料 | Marinade

魚露 1 湯匙
生粉 1 茶匙

1 tbsp fish sauce
1 tsp caltrop starch

## 調味料 | Seasoning

茄汁 1/2 杯
生粉 1 湯匙
糖 2 茶匙
雞粉 1 茶匙
鹽 1/2 茶匙
水 1 杯

1/2 cup ketchup
1 tbsp caltrop starch
2 tsps sugar
1 tsp chicken powder
1/2 tsp salt
1 cup water

## 做法 | Method

1. 雞肉洗淨,瀝乾水份,切粗粒,加入醃料拌勻。
2. 青、紅椒洗淨,去籽,切角;洋葱洗淨,去衣切粒。
3. 燒熱油鑊,爆香蒜茸、乾葱茸、洋葱、咖喱,加入雞肉、青、紅椒、葡萄乾,下調味料煮片刻即成。

1. Rinse chicken and drain. Cut into thick dice and mix with marinade.
2. Rinse green and red bell peppers. Seed and cut into wedges. Rinse onion. Skin and cut into dice.
3. Heat oil in a wok. Stir-fry grated garlic, grated shallots, onion and curry until fragrant. Add chicken, green and red bell peppers and raisins. Put in seasoning and cook for a while. Serve.

### 小貼士 | Tips

可煮多些咖喱雞醬放雪櫃中,隨時加熱,配飯、意粉、麵同吃。

# 青咖喱雞翼

## Simmered Chicken Wings with Green Curry

### 材料 | Ingredients

| | |
|---|---|
| 雞翼 400 克 | 400g chicken wings |
| 紅蘿蔔 1 個 | 1 carrot |
| 洋葱 1 個 | 1 onion |
| 茄子 100 克 | 100g eggplant |
| 九層塔 10 片 | 10 sweet basil leaves |
| 檸檬葉 2 片 | 2 Kaffir lime leaves |
| 椰汁 400 毫升 | 400ml coconut milk |
| 青咖喱醬 2 湯匙 | 2 tbsps green curry paste |
| 乾葱 3 粒 | 3 shallots |
| 水 2 杯 | 2 cups water |

⟨⟨⟩⟩ 醃料 | Marinade

鹽 1 茶匙
酒 1 茶匙
大蒜粉 1 茶匙

1 tsp salt
1 tsp wine
1 tsp garlic powder

⟨⟨⟩⟩ 調味料 | Seasoning

魚露 1 湯匙
糖 1 茶匙

1 tbsp fish sauce
1 tsp sugar

⟨⟨⟩⟩ 做法 | Method

1. 紅蘿蔔洗淨，去皮切塊；洋葱洗淨，去衣切塊。

2. 茄子洗淨，切塊；乾葱洗淨，去衣；九層塔洗淨。

3. 雞翼以醃料醃 30 分鐘，燒熱油鑊，下雞翼煎至兩面金黃色。

4. 燒熱油鑊，爆香乾葱、咖喱、洋葱，加入紅蘿蔔、茄子、檸檬葉、調味料和水，將雞翼回鑊，炆至汁稍乾，加入九層塔快手炒勻即可。

1. Rinse carrot. Peel and cut into pieces. Rinse onion. Skin and cut into pieces.

2. Rinse eggplant and cut into pieces. Rinse shallots and skin. Rinse sweet basil leaves.

3. Marinate chicken wings for 30 minutes. Heat oil in a wok. Fry the chicken wings until both sides are golden brown.

4. Heat oil in a wok. Stir-fry shallots, curry and onion until fragrant. Add carrot, eggplant, Kaffir lime leaves, seasoning and water. Put in the chicken wings and simmer until the sauce starts to dry. Add sweet basil leaves and stir-fry quickly until well-incorporated. Serve.

小貼士 | Tips

九層塔又名羅勒或金不換，常見於西式食譜及泰國菜。

# 椰香咖喱鴨

Simmered Duck with Curry and Coconut Milk

## 材料 | Ingredients

鴨件 400 克
乾葱 4 粒
蒜頭 2 粒
咖喱粉 2 茶匙

400g duck (chopped into pieces)
4 shallots
2 cloves garlic
2 tsps curry powder

2~4 人
2~4 persons

40~45 分鐘
40~45 minutesa

⊙⊙⊙ 醃料 | Marinade

酒 1 湯匙
生抽 2 茶匙
百里香 1 茶匙
胡椒粉 1/2 茶匙
小茴香粉 1/2 茶匙

1 tbsp wine
2 tsps light soy sauce
1 tsp thyme
1/2 tsp ground white pepper
1/2 tsp ground fennel

⊙⊙⊙ 調味料 | Seasoning

丁香 2 粒
椰汁 400 毫升
酸子 3 粒（用熱水 1/2 杯煮成汁）
糖 1 1/2 茶匙
鹽 3/4 茶匙
水 2 杯

2 cloves
400ml coconut milk
3 tamarinds (cooked with
1/2 cup of hot water into juice)
1 1/2 tsps sugar
3/4 tsp salt
2 cups water

⊙⊙⊙ 做法 | Method

1. 鴨件洗淨，加醃料醃 1 小時。
2. 燒熱油鑊，下鴨件泡油，盛起。
3. 再燒熱油鑊，爆香咖喱、乾葱、蒜頭，加入鴨件炒勻。
4. 下調味料（加入一半椰汁），以慢火炆至汁稍收乾，即加入餘下的椰汁略炒即成。

1. Rinse duck and marinate for 1 hour.
2. Heat oil in a wok. Stir-fry duck briefly and drain.
3. Heat oil in a wok again. Stir-fry curry, shallots and garlic quickly until fragrant. Add duck and stir-fry well.
4. Put in seasoning (add only half amount of coconut milk). Simmer over low heat until the sauce starts to dry. Add the remaining coconut milk and stir-fry briefly. Serve.

小貼士 | Tips

如用全隻鴨，一定要切去鴨尾的白色粒狀物體，才不會有羶味。

印尼椰汁咖喱牛
肉伴印度薄餅

Simmered Beef with Coconut
Milk and Curry in Indonesia
Style with Indian Chapati

2~4 人
2~4 persons

50~55 分鐘
50~55 minutesa

# 印尼薄餅
## Indian Chapati

### 小貼士 | Tips

搓麵粉時在工作
枱上灑上麵粉，
以避免薄餅黏在
一起。麵粉不能
搓得太久，否則
會變韌。

## ⦿⦿⦿ 材料 | Ingredients

| | |
|---|---|
| 麵粉 250 克 | 250g flour |
| 溶牛油 2 湯匙 | 2 tbsps molten butter |
| 鹽 1 茶匙 | 1 tsp salt |
| 水約 150 毫升 | approx. 150ml water |

## ⦿⦿⦿ 做法 | Method

1. 將所有材料混合，搓成粉糰。
2. 用布包裹，置室溫 40 分鐘待發。然後用麵粉棍將發好的粉糰輾成 5 毫米厚的圓形薄餅。
3. 燒熱油鑊，放下薄餅，以小火煎至薄餅脹起，翻轉另一面，煎至微金黃色即可拌咖喱享用。

1. Mix all ingredients and knead into a dough.
2. Wrap it with a cloth and set aside in room temperature for 40 minutes for fermentation. Roll the fermented dough flat with a roller into 5 mm thick round biscuits.
3. Heat oil in a wok. Fry the biscuits over low heat until expand. Turn upside down and fry until slightly golden. Serve with curry.

# 印尼椰汁咖喱牛肉

## Simmered Beef with Coconut Milk and Curry in Indonesia Style

### 材料 | Ingredients

| | |
|---|---|
| 牛柳 450 克 | 450g beef tenderloin |
| 椰汁 600 毫升 | 600ml coconut milk |

### 汁料 | Sauce

乾葱 10 粒
蒜頭 5 粒
石栗 3 粒
水 1/4 杯

10 shallots
5 cloves garlic
3 candlenuts
1/4 cup water

### 調味料 | Seasoning

| | |
|---|---|
| 咖喱粉 2 茶匙 | 2 tsps curry powder |
| 鹽 2 茶匙 | 2 tsps salt |
| 糖 1 茶匙 | 1 tsp sugar |
| 芫荽粉 1 茶匙 | 1 tsp ground coriander |
| 小茴香粉 1/2 茶匙 | 1/2 tsp ground fennel |
| 胡椒粉 1/8 茶匙 | 1/8 tsp ground white pepper |
| 南薑 3 片 | 3 slices galangal |
| 沙冧葉 2 片 | 2 salam leaves |

### 做法 | Method

1. 牛柳洗淨，切粒，放滾水中燙片刻後盛起。
2. 乾葱、蒜頭、石栗分別去衣，洗淨，加水放攪拌機內磨成茸，隔去粗粒成汁料。
3. 將汁料煮滾，注入椰汁 400 毫升和調味料煮滾。
4. 加入牛柳，慢火炆至汁料稍乾，再加入餘下的椰汁煮片刻，棄去南薑便可。

1. Rinse beef tenderloin and dice. Blanch it in boiling water for a while and drain.
2. Skin shallots, garlic and candlenuts respectively. Rinse and blend them with water in a blender until it is in mashed form. Drain away any granulates to become the sauce.
3. Bring the sauce to the boil. Add 400 ml of coconut milk and the seasoning. Bring to the boil.
4. Put in beef tenderloin. Simmer over low heat until the sauce starts to dry. Add the remaining coconut milk and cook for a while. Remove galangal and serve.

# 咖喱燜牛腩
## Simmered Beef Briskets with Curry

### 材料 | Ingredients

牛腩 500 克
薯仔 2 個
薑 10 片
香葉 1 片
蒜頭 3 粒

500g beef briskets
2 potatoes
10 slices ginger
1 slice bay leaf
3 cloves garlic

6~8 人
6~8 persons

20~25 分鐘
20~25 minutes

## 調味料 | Seasoning

| | |
|---|---|
| 咖喱醬 2 湯匙 | 2 tbsps curry paste |
| 咖喱粉 2 茶匙 | 2 tsps curry powder |
| 麵豉醬 2 茶匙 | 2 tsps fermented soybean paste |
| 冰糖 1 小塊 | 1 small cube rock sugar |

## 做法 | Method

1. 牛腩洗淨，汆水後切塊。
2. 薯仔洗淨，去皮切塊，下油鑊泡油備用。
3. 蒜頭洗淨，去衣切片。
4. 燒熱油鑊，爆香薑片、蒜片，加入牛腩、香葉、水，燜 40 分鐘，熄火焗 15 分鐘。
5. 另起油鑊，爆香咖喱醬、咖喱粉和麵豉醬，加入牛腩、薯仔、冰糖一同燜 20 分鐘即成。

1. Rinse beef briskets. Scald and cut into pieces.
2. Rinse potatoes. Peel and cut into pieces. Scald them in oil slowly and set aside.
3. Rinse garlic. Skin and slice.
4. Heat oil in a wok. Stir-fry ginger slices and garlic slices until fragrant. Add beef briskets, bay leaf and water. Simmer for 40 minutes. Remove heat and set aside with the lid covered for 15 minutes.
5. Heat oil in another wok. Stir-fry curry paste, curry powder and fermented soybean paste until fragrant. Add the beef briskets, potatoes and rock sugar. Simmer for 20 minutes and serve.

### 小貼士 | Tips

燜牛腩時加入冰糖會縮短烹調的時間。

咖喱肥牛粉絲煲

Curry Fat Beef and Vermicelli in Clay Pot

## 材料 | Ingredients

| | |
|---|---|
| 肥牛肉 300 克 | 300g fat beef |
| 粉絲 100 克 | 100g vermicelli |
| 指天椒 1 隻 | 1 bird's eye chilli |
| 咖喱醬 2 湯匙 | 2 tbsps curry paste |
| 蒜茸 1 湯匙 | 1 tbsp grated garlic |
| 薑茸 1 湯匙 | 1 tbsp grated ginger |
| 乾葱茸 1 茶匙 | 1 tsp grated shallot |
| 咖喱粉 1 茶匙 | 1 tsp curry powder |

醃料 | Marinade

生抽 2 茶匙
糖 1/2 茶匙
胡椒粉少許

2 tsps light soy sauce
1/2 tsp sugar
ground white pepper

調味料 | Seasoning

蠔油 1 湯匙
糖 1 茶匙
水適量

1 tbsp oyster sauce
1 tsp sugar
water

做法 | Method

1. 肥牛肉洗淨，加醃料醃20分鐘。燒熱油鑊，下牛肉泡油備用。
2. 粉絲浸軟。指天椒洗淨，切圈。
3. 燒熱砂窩，下油爆香蒜茸、乾葱茸、指天椒和咖喱略炒勻。
4. 加入粉絲，拌勻後加入調味料，放入牛肉，加蓋煮約 5 分鐘即成。

1. Rinse beef and marinate for 20 minutes. Heat oil in a wok. Scald beef in oil and set aside.
2. Soak vermicelli until soft. Rinse bird's eye chilli and cut into rings.
3. Heat oil in a clay pot. Stir-fry grated garlic, grated shallot, bird's eye chilli and curry briefly until well-incorporated.
4. Add vermicelli and mix well. Put in seasoning and beef. Cover the lid and cook for about 5 minutes. Serve.

小貼士 | Tips

醃牛肉時不要加鹽，否則會令肉質變韌。

# 泰式紅咖喱牛肋條

## Thai Red Curry Beef Short Ribs

### 材料 | Ingredients

| | |
|---|---|
| 牛肋條 500 克 | 500g beef short ribs |
| 番茄 2 個 | 2 tomatoes |
| 九層塔 10 片 | 10 sweet basil leaves |
| 乾葱 4 粒 | 4 shallots |
| 辣椒膏 1 湯匙 | 1 tbsp chilli paste |
| 紅咖喱粉 3 茶匙 | 3 tsps red curry powder |

4~6 人
4~6 persons

1 小時 20 分鐘
1 hour 20 minutes

**⦿⦿⦿ 調味料 | Seasoning**

| | |
|---|---|
| 魚露 1 湯匙 | 1 tbsp fish sauce |
| 糖 1 茶匙 | 1 tsp sugar |
| 胡椒粉 1 茶匙 | 1 tsp ground white pepper |
| 大蒜粉 1 茶匙 | 1 tsp garlic powder |
| 鹽 1/3 茶匙 | 1/3 tsp salt |

**⦿⦿⦿ 做法 | Method**

1. 牛肋條洗淨，汆水，切小段。
2. 番茄洗淨，切片。九層塔洗淨。
3. 燒熱油鑊，爆香乾葱、咖喱、辣椒膏，加入牛肋條、調味料，加水慢火燜煮 1 小時。
4. 加入番茄，再燜 10 分鐘，撒下九層塔即可。

1. Rinse beef short ribs. Scald and cut into small sections.
2. Rinse tomatoes and slice. Rinse sweet basil leaves.
3. Heat oil in a wok. Stir-fry shallots, curry and chilli paste until fragrant. Add beef short ribs and seasoning. Put in water and simmer over low heat for 1 hour.
4. Add tomatoes and simmer for 10 more minutes. Sprinkle over sweet basil leaves and serve.

**小貼士 | Tips**

如覺得咖喱辣味太重，可加入椰漿，以減少咖喱的辣味和令咖喱汁更香滑。

4~6 人
4~6 persons

15~20 分鐘
15~20 minutes

## 椰香咖喱燉酸辣牛肉

Sour and Spicy Curry Beef with Coconut Milk

### 材料 | Ingredients

牛肉 500 克
椰汁 100 毫升
咖喱醬 3 湯匙
咖喱粉 2 茶匙
乾葱 3 粒

500g beef
100 ml coconut milk
3 tbsps curry paste
2 tsps curry powder
3 shallots

### 醃料 | Marinade

生抽 2 茶匙
糖 1/2 茶匙
水 1 湯匙

2 tsps light soy sauce
1/2 tsp sugar
1 tbsp water

### 調味料 | Seasoning

酸子 4 粒（用水 1/2 杯煮至出味）
辣椒膏 2 湯匙
芫荽粉 1 茶匙
黃薑粉 1 茶匙

4 tamarinds (cooked with
1/2 cup of water until flavourful)
2 tbsps chilli paste
1 tsp coriander powder
1 tsp turmeric powder

### 做法 | Method

1. 牛肉洗淨，切成條狀，加醃料醃片刻；下油鑊泡油備用。
2. 乾葱洗淨，去衣。
3. 燒熱油鑊，爆香乾葱、咖喱，加入椰汁、調味料，將牛肉回鑊炒片刻即成。

1. Rinse beef and cut into strips. Marinate for a while. Scald beef in oil and set aside.
2. Rinse shallots and skin.
3. Heat oil in a wok. Stir-fry shallots and curry until fragrant. Add coconut milk and seasoning. Put in the beef and stir-fry for a while. Serve.

### 小貼士 | Tips

醃牛肉時加少許水，可令牛肉更嫩滑。

# 日式咖喱煎豬扒

Fried Pork Chop with Curry
in Japanese Style

## 材料 | Ingredients

豬扒 4 件
洋葱 2 個
薯仔 1 個
甘筍 1/2 條
日式咖喱磚 100 克

4 pork chops
2 onions
1 potato
1/2 carrot
100g Japanese curry cube

2~4 人
2~4 persons

20~25 分鐘
20~25 minutes

## 醃料 | Marinade

七味粉 1 茶匙
鹽 1/2 茶匙
胡椒粉 1/2 茶匙

1 tsp shichimi
(seven spices powder)
1/2 tsp salt
1/2 tsp ground white pepper

## 調味料 | Seasoning

生抽 1 茶匙
糖 1 1/2 茶匙
鹽 1/2 茶匙
水 1 杯

1 tsp light soy sauce
1 1/2 tsps sugar
1/2 tsp salt
1 cup water

## 做法 | Method

1. 豬扒洗淨，瀝乾水分，以醃料醃 1 小時。
2. 燒熱油鑊，將豬扒煎至金黃備用。
3. 洋葱、甘筍、薯仔洗淨，去衣或去皮，切塊。
4. 薯仔下沸水中煮腍後壓成茸。
5. 下油爆香洋葱、甘筍，將豬扒回鑊，加入咖喱磚輕輕壓熔，下調味料煮滾後收慢火，加入薯茸，煮至濃稠即可。

1. Rinse pork chops and drain. Marinate for 1 hour.
2. Heat oil in a wok. Fry the pork chops until golden brown and set aside.
3. Rinse onions, carrot and potato. Skin or peel respectively. Cut them into pieces.
4. Cook potato in boiling water until tender and mash.
5. Heat oil in a wok. Stir-fry onions and carrot until fragrant. Transfer the pork chops back into the wok. Add the curry cube and press slightly until molten. Put in the seasoning and bring to the boil. Reduce to low heat and add the mashed potato. Cook until the sauce thickens. Serve.

> **小貼士 | Tips**
> 薯仔壓茸放入汁內，可令醬汁更加濃郁。

4~6 人
4~6 persons

15~20 分鐘
15~20 minutes

緬甸咖喱豬扒

Curry Pork Chop in Burmaese Style

### 材料 | Ingredients

豬扒 500 克
乾葱 50 克
香茅 2 枝
咖喱醬 1 1/2 湯匙
咖喱粉 2 茶匙
大蒜粉 2 茶匙
水 2 杯

500g pork chops
50g shallots
2 stalks lemongrass
1 1/2 tbsps curry paste
2 tsps curry powder
2 tsps garlic powder
2 cups water

### 醃料 | Marinade

酒 1 湯匙
鹽 1 茶匙
糖 1 茶匙

1 tbsp wine
1 tsp salt
1 tsp sugar

### 小貼士 | Tips

豬扒如醃過夜會更入味。

### 做法 | Method

1. 豬扒洗淨，瀝乾水分，以刀背剁鬆，加入醃料醃約 30 分鐘。
2. 香茅洗淨，切段；乾葱洗淨，去衣。
3. 燒熱油鑊，將豬扒煎至六七成熟，盛起。
4. 下油爆香咖喱醬、咖喱粉、大蒜粉、乾葱、香茅，將豬扒回鑊，加水燜煮片刻即成。

1. Rinse pork chops and drain. Tap lightly with the back of a knife. Marinate for about 30 minutes.
2. Rinse lemongrass and cut into sections. Rinse shallots and skin.
3. Heat oil in a wok. Fry pork chops until medium-cooked and dish up.
4. Heat oil in a wok. Stir-fry curry paste, curry powder, garlic powder, shallots and lemongrass until fragrant. Add the pork chops. Add water and simmer for a while and serve.

# 紅咖喱排骨

## 材料 | Ingredients

排骨 300 克
青椒 1 個
紅辣椒 2 隻
乾葱 8 粒
咖喱醬 1 湯匙
紅咖喱粉 3 茶匙

300g spareribs
1 green bell pepper
2 red chillies
8 shallots
1 tbsp curry paste
3 tsps red curry powder

1~2 人
1~2 persons

15~20 分鐘
15~20 minutes

## 醃料 | Marinade

生抽 1 茶匙
胡椒粉 1 茶匙
生粉 1 茶匙
1 tsp light soy sauce
1 tsp ground white pepper
1 tsp caltrop starch

## 調味料 | Seasoning

生抽 1 茶匙
麻油 1 茶匙
老抽 1/2 茶匙
水 5 湯匙
1 tsp light soy sauce
1 tsp sesame oil
1/2 tsp dark soy sauce
5 tbsps water

## 芡汁 | Thickening

生粉 1 茶匙
水 3 湯匙
1 tsp corn starch
3 tbsps water

### 小貼士 | Tips
排骨用刀拍鬆後再醃，
可令肉質更軟滑。

## 做法 | Method

1. 青椒洗淨，去籽，切角。
2. 紅辣椒洗淨，切圈；乾葱洗淨，去衣。
3. 排骨洗淨，加醃料醃 30 分鐘。
4. 燒熱油鑊，爆香乾葱、紅辣椒、咖喱，加入排骨和調味料煮至排骨熟透，加入青椒拌勻，下生粉水勾芡即可上碟。

1. Rinse green bell pepper. Seed and cut into wedges.
2. Rinse red chillies and cut into rings. Rinse shallots and skin.
3. Rinse spareribs and marinate for 30 minutes.
4. Heat oil in a wok. Stir-fry shallots, red chillies and curry until fragrant. Add spareribs and seasoning. Cook until the spareribs are done. Mix in green bell pepper. Thicken the sauce with thickening. Serve.

## 印度咖喱羊肉

### Indian Curry Mutton

◯◯◯ 材料 | Ingredients

| | |
|---|---|
| 羊肉 500 克 | 500g mutton |
| 番茄 2 個 | 2 tomatoes |
| 薯仔 2 個 | 2 potatoes |
| 紅辣椒 2 隻 | 2 red chillies |
| 乾葱 8 粒 | 8 shallots |
| 咖喱醬 2 湯匙 | 2 tbsps curry paste |
| 咖喱粉 2 茶匙 | 2 tsps curry powder |

◯◯◯ 醃料 | Marinade

| | |
|---|---|
| 酒 1 湯匙 | 1 tbsp wine |
| 胡椒粉 1 茶匙 | 1 tsp ground white pepper |
| 鹽 1 茶匙 | 1 tsp salt |
| 百里香少許 | thyme |

## 調味料 | Seasoning

| | |
|---|---|
| 丁香 2 粒 | 2 cloves |
| 八角 1 粒 | 1 star aniseed |
| 魚露 1 湯匙 | 1 tbsp fish sauce |
| 黃薑粉 2 茶匙 | 2 tsps turmeric powder |
| 肉桂粉 1 茶匙 | 1 tsp ground cinnamon |
| 豆蔻粉 1 茶匙 | 1 tsp ground nutmeg |
| 大蒜粉 1 茶匙 | 1 tsp garlic powder |
| 水 6 杯 | 6 cups water |

## 做法 | Method

1. 羊肉洗淨，切件，加醃料醃 30 分鐘。
2. 番茄洗淨，切塊；薯仔洗淨，去皮切塊。
3. 乾葱洗淨，去衣切碎；紅辣椒洗淨，切丁。
4. 燒熱油鑊，爆香咖喱、乾葱、紅辣椒，下羊肉、薯仔，加入調味料和水，燜煮 1 小時 30 分鐘，再加入番茄煮 10 分鐘，下鹽炒勻即可。

1. Rinse mutton and chop into pieces, marinate for 30 minutes.
2. Rinse tomatoes and cut into pieces. Rinse potatoes. Peel and cut into pieces.
3. Rinse shallots. Skin and chop. Rinse chillies and cut into small cubes.
4. Heat oil in a wok. Stir-fry curry, shallots and red chillies until fragrant. Add mutton and potatoes. Put in seasoning and water. Simmer for 1 hour 30 minutes. Add tomatoes and cook for 10 minutes. Put in salt and stir-fry well. Serve.

---

小貼士 | Tips

下了豆蔻粉可令汁料更濃稠和香滑。

## 酸子咖喱燴魚

### Braised Curry Fish with Tamarinds

◯◯ 材料 | Ingredients

| | |
|---|---|
| 鯉魚 1 條 | 1 carp |
| 番茄 4 個（去皮） | 4 tomatoes (skinned) |
| 青椒 1 個 | 1 green bell pepper |
| 紅燈籠椒 1 個 | 1 red bell pepper |
| 洋葱 1/2 個 | 1/2 onion |
| 蒜茸 3 粒量 | 3 cloves garlic (grated) |
| 香茅 1 枝 | 1 stalk lemongrass |
| 薑 5 片 | 5 slices ginger |
| 咖喱醬 2 湯匙 | 2 tbsps curry paste |
| 辣椒膏 1 湯匙 | 1 tbsp chilli paste |

2~4 人
2~4 persons

20~25 分鐘
20~25 minutes

## 調味料 | Seasoning

| | |
|---|---|
| 酸子 6 粒（用 1/2 杯水煮成酸汁） | 6 tamarinds (cooked with |
| 魚露 1 湯匙 | 1/2 cup of water into juice) |
| 糖 2 茶匙 | 1 tbsp fish sauce |
| | 2 tsps sugar |

## 做法 | Method

1. 魚劏洗淨，瀝乾水分。
2. 青、紅椒洗淨，去籽，切粒；洋葱洗淨，去衣，切粒；香茅洗淨，切段。
3. 番茄洗淨，用刀在底部剠十字，去皮後切粒。
4. 魚撲上生粉，燒熱油鑊，將鮮魚炸至金黃色。
5. 下油爆香蒜茸、香茅、薑、咖喱、辣椒膏，加入所有切粒材料、調味料，煮至汁料稍收乾，淋在魚上即成。

1. Gut and rinse the fish and drain.
2. Rinse green and red bell peppers. Seed and cut them into dice. Rinse onion. Skin and cut into dice. Rinse lemongrass and cut into sections.
3. Rinse tomatoes and slit a cross at the bottom. Skin and cut into dice.
4. Coat the fish with caltrop starch. Heat oil in a wok. Deep-fry the fish until golden brown.
5. Heat oil in a wok. Stir-fry grated garlic, lemongrass, ginger, curry and chilli paste until fragrant. Add all he diced ingredients and seasoning. Cook until the sauce starts to dry and pour over the fish. Serve.

### 小貼士 | Tips

炸魚前先在魚身斜剠兩刀，可令魚更快熟和容易吸收汁料。

# 咖喱魚頭煲

## Curry Fish Head in Clay Pot

### 材料 | Ingredients

| | |
|---|---|
| 大魚頭 1 個 | 1 big fish head |
| 番茄 2 個 | 2 tomatoes |
| 南薑 4 片 | 4 slices galangal |
| 洋葱 1 個 | 1 onion |
| 乾葱 5 粒 | 5 shallots |
| 香茅 2 枝 | 2 stalks lemongrass |
| 葱 3 棵 | 3 stalks spring onion |
| 酸子 5 粒 | 5 tamarinds (cooked with |
| （用 1/2 杯水煮成酸汁） | 1/2 cup of water into juice) |
| 咖喱醬 2 湯匙 | 2 tbsps curry paste |
| 辣椒膏 2 茶匙 | 2 tsps chilli paste |
| 蝦膏 1 茶匙 | 1 tsp shrimp paste |
| 水 1 杯 | 1 cup water |
| 鹽適量 | salt |

## 醃料 | Marinade

酒 2 茶匙
胡椒粉 1 茶匙
鹽 1 茶匙

2 tsps wine
1 tsp ground white pepper
1 tsp salt

## 調味料 | Seasoning

酸子汁 3 湯匙
魚露 2 茶匙
糖 1 茶匙

3 tbsps tamarind juice
2 tsps fish sauce
1 tsp sugar

## 做法 | Method

1. 魚頭加醃料醃 15 分鐘。
2. 番茄洗淨，切塊；洋葱洗淨，去衣切絲；南薑洗淨。
3. 乾葱洗淨，去衣切片；香茅、葱洗淨，切段。
4. 燒熱油鑊，下魚頭煎至金黃備用。
5. 下油爆香香茅、乾葱、咖喱、辣椒膏、蝦膏、洋葱，加入番茄、南薑、水、調味料煮 15 分鐘，加入魚頭和鹽煮片刻，加葱段拌勻即成。

1. Marinate the fish head for 15 minutes.
2. Rinse tomatoes and cut into pieces. Rinse onion. Skin and cut into shreds. Rinse galangal.
3. Rinse shallots. Skin and cut into slices. Rinse lemongrass and spring onion. Cut them into sections.
4. Heat oil in a wok. Fry the fish head until golden brown and set aside.
5. Heat oil in a wok. Stir-fry lemongrass, shallots, curry, chilli paste, shrimp paste and onion until fragrant. Add tomatoes, galangal, water and seasoning. Cook for 15 minutes. Put in the fish head and salt. Cook for a while. Mix in spring onion and serve.

小貼士 | Tips

魚頭加入酒來醃，可去除腥味。

# 泰式咖喱魚頭

## Thai Curry Fish Head

### 材料 | Ingredients

| | |
|---|---|
| 三文魚頭 1 個 | 1 salmon fish head |
| 免治豬肉 100 克 | 100g minced pork |
| 芹菜 2 棵 | 2 stalks Chinese celery |
| 香茅 1 枝 | 1 stalk lemongrass |
| 紅辣椒 2 隻 | 2 red chillies |
| 咖喱醬 2 湯匙 | 2 tbsps curry paste |
| 花生醬 2 湯匙 | 2 tbsps peanut butter |
| 辣椒膏 1 湯匙 | 1 tbsp chilli paste |
| 乾葱茸 2 粒量 | 2 shallots (grated) |
| 蒜茸 2 粒量 | 2 cloves garlic (grated) |
| 雞湯 380 毫升 | 380ml chicken broth |
| 檸檬葉 | lime leaves |

### 豬肉醃料 | Marinade for pork

| | |
|---|---|
| 生粉 1 茶匙 | 1 tsp caltrop starch |
| 鹽 1/2 茶匙 | 1/2 tsp salt |

4~6 人
4~6 persons

45~50 分鐘
45~50 minutes

## 魚頭醃料 | Marinade for fish head

生粉 1 茶匙
酒 1 茶匙
鹽 1/2 茶匙

1 tsp caltrop starch
1 tsp wine
1/2 tsp salt

## 汁料 | Sauce

魚露 1 茶匙
青檸汁 1/2 個或 1 個量

1 tsp fish sauce
1/2 or 1 lime (juiced)

## 做法 | Method

1. 免治豬肉以豬肉醃料醃 20 分鐘。

2. 芹菜、香茅洗淨，切度。

3. 三文魚頭洗淨，瀝乾水分，以魚頭醃料醃 15 分鐘。

4. 燒熱油鑊，下三文魚頭炸至金黃色，盛起備用。

5. 將雞湯放煲內煮沸，加入花生醬拌勻。

6. 另起油鑊，爆香咖喱、辣椒膏、香茅、蒜茸、乾葱茸，加入檸檬葉、豬肉、花生醬和紅辣椒，放雞湯煲內，加入汁料，煮滾後放下魚頭和芹菜，煮滾片刻即可。

> **小貼士 | Tips**
> 芹菜又名唐芹或香芹，可分為水芹和旱芹；其中以水芹食味較佳。

1. Marinate minced pork for 20 minutes.

2. Rinse Chinese celery and lemongrass. Cut into sections.

3. Rinse salmon fish head and drain, marinate for 15 minutes.

4. Heat oil in a wok. Deep-fry the salmon fish head until golden brown and drain.

5. Bring chicken broth to the boil in a pot. Add peanut butter and mix well.

6. Heat oil in another wok. Stir-fry curry, chilli paste, lemongrass, grated garlic and grated shallots until fragrant. Add in lime leaves, pork, peanut butter and red chillies. Transfer the mixture into the chicken broth. Put in the sauce. Bring to the boil and add the fish head and Chinese celery. Boil for a while and serve.

## 香茅咖喱蟹

### Lemongrass and Curry Crabs

材料 | Ingredients

肉蟹 4 隻
4 mud crabs

醃料 | Marinade

鹽少許
salt

## 脆漿料 | Deep-frying paste

| | |
|---|---|
| 麵粉 1/2 杯 | 1/2 cup flour |
| 油 1 湯匙 | 1 tbsp oil |
| 咖喱粉 2 茶匙 | 2 tsps curry powder |
| 香茅粉 2 茶匙 | 2 tsps lemongrass powder |
| 大蒜粉 1 茶匙 | 1 tsp garlic powder |
| 椒鹽 1/2 茶匙 | 1/2 tsp peppery salt |
| 水 1/2 茶匙 | 1/2 tsp water |

## 做法 | Method

1. 麵粉加入咖喱粉、香茅粉、大蒜粉、椒鹽拌勻，加水調成漿狀，面放油即成脆漿，放雪櫃待用。
2. 肉蟹劏洗淨，吸乾水分，加少許鹽略醃。
3. 將蟹切件，沾上脆漿；燒熱油鑊，下油中炸至金黃色即成。

1. Add curry powder, lemongrass powder, garlic powder and peppery salt into the flour and mix well. Add in water and mix into paste form. Add oil on top to make the deep-frying paste. Set aside in the refrigerator.
2. Gut crabs and rinse. Wipe dry and marinate briefly with a little salt.
3. Cut the crabs into pieces and coat with the deep-frying paste. Heat oil in a wok. Deep-fry until golden brown. Serve.

小貼士 | Tips

天氣熱時，脆漿一定要放雪櫃，天氣比較涼時則可放在陰涼處。

# 咖喱粉絲蟹煲

## 材料 | Ingredients

| | |
|---|---|
| 肉蟹 3 隻 | 3 mud crabs |
| 粉絲 100 克 | 100g vermicelli |
| 紅燈籠椒 1 隻 | 1 red bell pepper |
| 薑 6 片 | 6 slices ginger |
| 葱 3 棵 | 3 stalks spring onion |
| 芫荽 1 棵 | 1 stalk coriander |
| 蒜頭 2 粒 | 2 cloves garlic |
| 上湯 1 1/2 杯 | 1 1/2 cups stock |
| 酒 1 湯匙 | 1 tbsp wine |
| 咖喱粉 2 茶匙 | 2 tsps curry powder |
| 麻油 1 茶匙 | 1 tsp sesame oil |
| 生粉適量 | caltrop starch |

4~6 人
4~6 persons

20~25 分鐘
20~25 minutes

## 調味料 | Seasoning

| | |
|---|---|
| 魚露 1 湯匙 | 1 tbsp fish sauce |
| 老抽 1 湯匙 | 1 tbsp dark soy sauce |
| 糖 1 茶匙 | 1 tsp sugar |
| 胡椒粉 1/2 茶匙 | 1/2 tsp ground white pepper |

## 做法 | Method

1. 蒜頭去衣，洗淨；紅燈籠椒洗淨，去籽，切角；葱、芫茜洗淨，切段。
2. 蟹劏好、洗淨，灑上生粉；燒熱油鑊，放油中略炸後盛起。
3. 粉絲洗淨，浸軟切段，放煲中，加入上湯。
4. 下油爆香蒜頭、薑、紅燈籠椒、咖喱粉及葱白，將蟹回鑊，下調味料，灒酒，拌勻，倒入粉絲煲內。
5. 加蓋燜約 10 分鐘，撒入其餘的葱段、芫茜，灑上麻油，即可原煲上桌。

1. Skin garlic and rinse. Rinse red bell pepper. Seed and cut into wedges. Rinse spring onion and coriander. Section.
2. Gut and rinse crabs. Sprinkle caltrop starch on top. Heat oil in a wok and deep-fry the crabs briefly.
3. Rinse vermicelli. Soak until soft and section. Put it into a pot and add stock.
4. Heat oil in a wok. Stir-fry garlic, ginger, red bell pepper, curry powder and white part of spring onion until fragrant. Return crabs and seasoning. Sizzle in wine. Mix well and transfer to the pot containing vermicelli.
5. Cover the lid of pot and simmer for about 10 minutes. Add the remaining spring onion and coriander. Pour over sesame oil. Serve.

## 小貼士 | Tips

蟹要即劏即炸才新鮮，此外，也可以水蟹代替。

# 黃金咖喱蟹

## Curry Crabs with Salted Egg Yolks

### 材料｜Ingredients

| | |
|---|---|
| 蟹 2 隻 | 2 crabs |
| 洋葱 1 個 | 1 onion |
| 薑 6 片 | 6 slices ginger |
| 香茅 2 枝 | 2 stalks lemongrass |
| 鹹蛋黃 6 隻 | 6 salted egg yolks |
| 紅辣椒 2 隻 | 2 red chillies |
| 葱 3 棵 | 3 stalks spring onion |

## 調味料 | Seasoning

| | |
|---|---|
| 咖喱醬 3 湯匙 | 3 tbsps curry paste |
| 魚露 1/2 湯匙 | 1/2 tbsp fish sauce |
| 南薑粉 1 茶匙 | 1 tsp ground galangal |
| 胡椒粉 1 茶匙 | 1 tsp ground white pepper |

## 做法 | Method

1. 蟹劏洗淨，切件。
2. 洋葱洗淨，去衣切片；紅辣椒洗淨，切絲；香茅、葱洗淨，切度。
3. 鹹蛋黃蒸熟，壓成茸備用。
4. 將蟹撲上生粉，起油鑊，下蟹件泡油，盛起備用。
5. 下油爆香洋葱、香茅、薑，下調味料，將蟹回鑊，撒下葱段，再盛起。
6. 下少許油，加入鹹蛋黃茸半推半炒至起泡，將蟹再次回鑊拌勻即可。

1. Gut and rinse crabs. Cut into pieces.
2. Rinse onion. Skin and cut into slices. Rinse red chillies and cut into shreds. Rinse lemongrass and spring onion. Cut into sections.
3. Steam salted egg yolks until done. Mash and set aside.
4. Coat crabs with caltrop starch lightly. Scald in oil and drain.
5. Heat oil in a wok. Stir-fry onion, lemongrass and ginger until fragrant. Add seasoning and the crabs. Stir-fry well. Drizzle over sectioned spring onion and dish up.
6. Heat a little oil in a wok. Add the mashed salted egg yolks. Stir-fry and push around until bubbles come out. Pour back the crabs and mix well. Serve.

小貼士 | Tips

可以酌量多加些鹹蛋黃，令此菜式更香，但膽固醇會更高，不利健康。

# 大蝦喇沙

## Prawns Laksa

### 材料 | Ingredients

| | |
|---|---|
| 大蝦（連殼）4 隻 | 4 large prawns (with shells) |
| 米線 400 克 | 400g rice noodles |
| 魚腐 4 個 | 4 fried fish puffs |
| 洋葱 1 個 | 1 onion |
| 香茅 3 枝 | 3 stalks lemongrass |
| 南薑 4 片 | 4 slices galangal |
| 芽菜 100 克 | 100g soybean sprouts |
| 葱 2 棵 | 2 stalks spring onion |
| 雞湯 2 杯 | 2 cups chicken broth |

2~4 人
2~4 persons

15~20 分鐘
15~20 minutes

◎◎ 調味料 | Seasoning

| | |
|---|---|
| 椰汁 200 毫升 | 200ml coconut milk |
| 咖喱粉 2 湯匙 | 2 tbsps curry powder |
| 辣椒膏 1 湯匙 | 1 tbsp chilli paste |
| 魚露 1 1/2 湯匙 | 1 1/2 tbsps fish sauce |
| 蝦醬 1 茶匙 | 1 tsp shrimp paste |
| 糖 1 茶匙 | 1 tsp sugar |
| 水 1 杯 | 1 cup water |

◎◎ 做法 | Method

1. 大蝦洗淨,瀝乾水分,剪去腳和鬚。
2. 魚腐、南薑、芽菜分別洗淨,瀝乾水分。
3. 洋葱洗淨,去衣切絲;香茅洗淨,切段;葱洗淨,切度。
4. 燒熱油鑊,爆香洋葱、香茅、南薑,下調味料和雞湯煮 15 分鐘。
5. 放入大蝦、魚腐、芽菜和葱,煮滾後加入米線拌勻即成。

1. Rinse large prawns and drain. Trim off legs and antennae.
2. Rinse fried fish puffs, galangal and soybean sprouts respectively. Drain.
3. Rinse onion. Skin and cut into shreds. Rinse lemongrass and section. Rinse spring onion and section.
4. Heat oil in a wok. Stir-fry onion, lemongrass and galangal until fragrant. Add seasoning and chicken broth. Cook for 15 minutes.
5. Put in large prawns, fried fish puffs, soybean sprouts and spring onion. Bring to the boil and mix in rice noodles. Serve.

小貼士 | Tips

洋葱開邊,放凍水內浸片刻才切,就不會刺激眼睛。

2~4 人
2~4 persons

15~20 分鐘
15~20 minutes

緬甸咖喱大蝦

Curry Prawns in Myanmar Style

## 材料 | Ingredients

| | |
|---|---|
| 大蝦 300 克 | 300g prawns |
| 洋葱 1 個 | 1 onion |
| 芹菜 3 棵 | 3 stalks Chinese celery |
| 葱 4 棵 | 4 stalks spring onion |
| 蒜茸 3 粒量 | 3 cloves garlic (grated) |
| 雞蛋 1 隻 | 1 egg |
| 咖喱醬 3 茶匙 | 3 tsps curry paste |
| 咖喱粉 2 茶匙 | 2 tsps curry powder |

## 調味料 | Seasoning

魚露 3 茶匙
糖 1 茶匙

3 tsps fish sauce
1 tsp sugar

### 小貼士 | Tips

蝦亦可不去殼，只剪去腳和鬚，全隻泡油。

## 做法 | Method

1. 蝦洗淨，去殼，留尾部分。

2. 燒熱油鑊，加入蝦泡油，盛起備用。

3. 洋葱洗淨，去衣切塊；芹菜、葱洗淨，切段；雞蛋打勻。

4. 下油爆香蒜茸、洋葱、咖喱、芹菜，將蝦回鑊，加雞蛋液，下調味料，以大火炒勻，撒下葱段即成。

1. Rinse prawns and shell. Keep the tails.

2. Heat oil in a wok. Scald prawns slowly in oil and drain.

3. Rinse onion. Skin and cut into pieces. Rinse Chinese celery and spring onion. Cut into sections. Whisk egg.

4. Heat oil in a wok. Stir-fry grated garlic, onion, curry and Chinese celery until fragrant. Pour back the prawns. Add whisked egg and the seasoning. Stir-fry over high heat until well-incorporated. Drizzle over sectioned spring onion and serve.

# 泰式咖喱鱿鱼

## Thai Curry Squids

### 材料 | Ingredients

| | |
|---|---|
| 鱿鱼 2 條 | 2 squids |
| 洋葱 1 個 | 1 onion |
| 青椒 1 個 | 1 green bell pepper |
| 紅辣椒 1 隻 | 1 red chilli |
| 番茄 1 個 | 1 tomato |
| 九層塔 10 片 | 10 basil leaves |
| 咖喱粉 3 湯匙 | 3 tbsps curry powder |

2~4 人
2~4 persons

15~20 分鐘
15~20 minutes

## 調味料 | Seasoning

| | |
|---|---|
| 魚露 2 茶匙 | 2 tsps fish sauce |
| 香茅粉 1 茶匙 | 1 tsp lemongrass powder |
| 黃薑粉 1 茶匙 | 1 tsp ground turmeric |
| 大蒜粉 1 茶匙 | 1 tsp garlic powder |
| 南薑粉 1 茶匙 | 1 tsp ground galangal |

## 做法 | Method

1. 魷魚劏洗淨，以斜刀剠花，切塊。
2. 洋葱洗淨，去衣切絲；青椒洗淨，去籽切絲；番茄洗淨，切丁。
3. 紅辣椒洗淨，切圈；九層塔洗淨。
4. 燒熱油鑊，爆香咖喱、洋葱、辣椒，加入番茄、青椒、調味料、魷魚炒至熟透，撒下九層塔快速炒勻即可。

1. Gut and rinse squids. Slit crosses on the internal side. Cut them into pieces.
2. Rinse onion. Skin and cut into shreds. Rinse green bell pepper. Seed and cut into shreds. Rinse tomato and cut into dice.
3. Rinse red chilli and cut into rings. Rinse basil leaves.
4. Heat oil in a wok. Stir-fry curry, onion and chilli until fragrant. Add tomato, green bell pepper, seasoning and squids. Stir-fry until done. Put in basil leaves and stir-fry quickly until incorporated. Serve.

### 小貼士 | Tips

剠魷魚花，是在魷魚的肚內側，炒熟時才會捲起。

## 葡汁焗鮮蔬

Baked Vegetables in Portuguese Style

◯◯◯ 材料 | Ingredients

| | |
|---|---|
| 西蘭花 100 克 | 100g broccoli |
| 椰菜花 100 克 | 100g cauliflower |
| 蘑菇 100 克 | 100g straw mushrooms |
| 珍珠筍 100 克 | 100g pearl shoots |
| 鮮蘆筍 1 條 | 1 fresh asparagus |
| 甘筍（細）1 條 | 1 small carrot |
| 牛油 50 克 | 50g butter |
| 麵粉 5 湯匙 | 5 tbsps flour |
| 洋葱茸 1 湯匙 | 1 tbsp chopped onion |

## 汁料 | Sauce

咖喱粉 1 茶匙
淡奶 180 毫升
椰汁 120 毫升
水 1 杯

1 tsp curry powder
180ml evaporated milk
120ml coconut milk
1 cup water

## 調味料 | Seasoning

鹽 2 茶匙
糖 1/2 茶匙

2 tsps salt
1/2 tsp sugar

### 小貼士 | Tips

假如沒有焗爐，可在鑊中煮至濃稠即可。

## 做法 | Method

1. 西蘭花、椰菜花洗淨，切成小朵；甘筍洗淨，去皮切塊；蘆筍洗淨，切段，以上各材料同汆水備用。

2. 蘑菇、珍珠筍洗淨，瀝乾水分。

3. 燒熱油鑊，下牛油炒香洋葱茸，加入麵粉炒勻。

4. 再加入汁料煮成糊狀，下調味料，倒入各材料拌勻，放在焗盆內，再放已預熱的焗爐內，以 150℃ 焗約 20 分鐘至金黃色即可。

1. Rinse broccoli and cauliflower. Cut into small florals. Rinse carrot. Peel and cut into pieces. Rinse asparagus and section. Scald the above ingredients together and set aside.

2. Rinse straw mushrooms and pearl shoots. Drain.

3. Heat oil in a wok. Stir-fry butter and chopped onion until fragrant. Add flour and stir-fry well.

4. Put in the sauce and cook into a paste. Add seasoning and all the ingredients. Mix well. Pour the mixture into a baking tray and bake in a preheated oven at 150°C for about 20 minutes until golden brown. Serve.

印度咖喱雜菜

Indian Curry Vegetables

2~4 人
2~4 persons

25~30 分鐘
25~30 minutes

## 材料 | Ingredients

番茄 3 個
洋蔥 1 個
青瓜 1 條
紅蘿蔔 1 條
小鹹魚乾 100 克

3 tomatoes
1 onion
1 cucumber
1 carrot
100g salted small fish

## 調味料 | Seasoning

咖喱粉 2 湯匙
辣椒膏 2 湯匙
魚露 1 湯匙
糖 1 湯匙

2 tbsps curry powder
2 tbsps chilli paste
1 tbsp fish sauce
1 tbsp sugar

## 做法 | Method

1. 洋蔥洗淨，去衣切塊；番茄洗淨，切塊；青瓜、紅蘿蔔洗淨，去皮切塊。
2. 鹹魚乾洗淨。
3. 燒熱油鑊，爆香洋蔥，加入番茄、青瓜、紅蘿蔔、鹹魚乾、調味料，以中火炆煮 15 分鐘即成。

1. Rinse onion. Skin and cut into pieces. Rinse tomatoes and cut into pieces. Rinse cucumber and carrot. Peel and cut them into pieces.
2. Rinse salted fish.
3. Heat oil in a wok. Stir-fry onion until fragrant. Add tomatoes, cucumber, carrot, salted fish and seasoning. Simmer over medium heat for 15 minutes. Serve.

### 小貼士 | Tips

材料中如有番茄，可以減少水的份量，因番茄會增加水分。

# 常用配料和香料

咖喱粉
curry powder

黃薑粉
ground turmeric

紅辣椒粉
red chilli powder

南薑粉
ground galangal

香茅粉
lemongrass powder

芫茜粉
ground coriander

小茴香粉
ground fennel

七味粉
shichimi
(seven spices powder)

咖喱醬 / 濕咖喱
curry paste
/ moist curry

青咖喱醬
green curry paste

泰式辣椒醬
Thai chilli paste

白豆蔻
nutmeg

小茴香
fennel

百里香
thyme

石栗
candlenuts

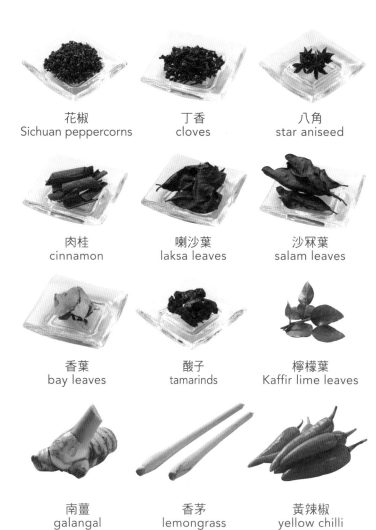

花椒
Sichuan peppercorns

丁香
cloves

八角
star aniseed

肉桂
cinnamon

喇沙葉
laksa leaves

沙冧葉
salam leaves

香葉
bay leaves

酸子
tamarinds

檸檬葉
Kaffir lime leaves

南薑
galangal

香茅
lemongrass

黃辣椒
yellow chilli

青胡椒
green pepper

惹味咖喱

**編著**
梁燕

**編輯**
Alvin、Pheona

**翻譯**
梁悅冰

**攝影**
Fanny

**美術設計**
Carol

**排版**
Rosemary

**出版者**
萬里機構出版有限公司
香港鰂魚涌英皇道1065號東達中心1305室
電話：2564 7511
傳真：2565 5539
電郵：info@wanlibk.com
網址：http://www.wanlibk.com
　　　http://www.facebook.com/wanlibk

**發行者**
香港聯合書刊物流有限公司
香港新界大埔汀麗路36號
中華商務印刷大廈3字樓
電話：2150 2100
傳真：2407 3062
電郵：info@suplogistics.com.hk

**承印者**
美雅印刷製本有限公司

**出版日期**
二零一九年二月第一次印刷

萬里機構

萬里 Facebook